CITIES OF THE FUTURE

Breakthrough Innovations That Will Change The Way We Live

A Beginner's Guide to Urban Innovations

Ram Khandelwal

DISCLAIMER
The information contained in this book is true to the best of our knowledge. However, any mistake due to human errors shall be omitted. The authors and publishers do not bear responsibility for any direct or indirect losses arising out of use of this book by the intended reader or any third party.

Copyright © 2023 Ram Khandelwal All rights reserved

No part of this book may be reproduced, or stored in a retrieval system, or transmitted in any form or by any means, electronic, mechanical, photocopying, recording, or otherwise, without express written permission of the publisher.

Cover design by: UrbanPIE
Print design by: Karthik Girish

Publisher: Urban Innovation Lab

DEDICATION

This book is dedicated to my beloved wife and son who are my greatest source of inspiration.

PREFACE

In this book, we explore the latest perspectives on urban innovation, and examine how cities around the world are using new technologies, policies, and approaches to address some of the most pressing challenges of our time. From new forms of citizen engagement and digital innovation to cutting-edge solutions for sustainable transportation and resilient infrastructure, the book provides a comprehensive look at the ways in which cities are driving progress and shaping the future.

We begin by examining the key drivers of urban innovation and the ways in which cities are leveraging new technologies and data to improve decision-making and service delivery. We then delve into specific case studies and examples of urban innovation from around the world, highlighting the successes and challenges of different approaches.

Throughout the book, we also explore the role of government and other key stakeholders in driving urban innovation, and the ways in which cities are engaging citizens and collaborating with the private sector and other partners to co-create new solutions.

We believe that this book will be an invaluable resource for city administrators, city managers, policymakers, urban planners and designers, architects, civil engineers, technology engineers and anyone interested in understanding the cutting-edge trends and strategies for creating more livable, sustainable, and resilient cities of the future.

CONTENTS

1

FUNDAMENTALS OF INNOVATION

Innovation distinguishes between a leader and a follower.

- Steve Jobs

Innovation is like when you come up with a new idea, like a new game to play or a new way to build a block tower. It's when you take something that already exists and make it better or come up with something completely new. Like when you take your toy car and add a ramp to make it go faster or when you build a castle out of blocks instead of a house. It's like using your imagination to make something even more fun and exciting.

What Is Innovation?

Innovation refers to the process of introducing new ideas, products, services, or processes into an organization or society. It can take many forms, such as developing a new product or improving an existing one, creating a new business model, or introducing a new way of working. Innovation is often seen as key to driving growth and competitiveness in today's fast-paced business environment. It can be incremental, which means small improvements to existing products or processes, or disruptive, which refers to new technologies or business models that fundamentally change the way an industry operates.

For instance, smartphones represented a disruptive innovation in the mobile phone industry, as they combined the functionality of a computer, camera, and GPS with traditional phone capabilities. This change in the way mobile phones were used, brought a lot of new features and services that were not available before, such as mobile internet access, apps and mobile payments. Additionally, smartphones have also led to the development of new industries, such as app development and mobile commerce. Smartphones have also transformed how people communicate and access information, changing the way we live and work in the process.

Why Innovation Is Important?

In today's fast-paced business environment, innovation

plays a crucial role in driving growth and competitiveness. Companies that are able to continuously innovate and adapt to changing market conditions are more likely to succeed in the long term. Innovation allows companies to differentiate themselves from their competitors by introducing new products, services, or processes. This can attract new customers and increase market share. Additionally, innovation can lead to efficiency improvements and cost savings, which can increase profitability.

Innovation helps companies to stay relevant in a rapidly changing market. For example, new technologies and trends can disrupt traditional business models, and companies that are able to adapt and adopt these changes are more likely to survive. Innovation also plays a key role in creating new markets and industries. Companies that are able to identify and capitalize on emerging trends and technologies can create new opportunities for growth and expansion.

Overall, innovation is essential for companies to stay competitive in today's fast-paced business environment and to continue to grow and evolve over time.

What Is The Difference Between Innovation And Invention?

Invention and innovation are related but distinct concepts. Invention refers to the creation of something new, such as a new product, technology, or process. It can be a physical object, like a machine, or an idea, like a new mathematical formula. Inventors often focus on solving a specific problem or improving upon existing solutions.

Innovation, on the other hand, refers to the successful implementation and commercialization of an invention. It is the process of taking an invention and bringing it to market, making it available to customers and creating value for the business or society. Innovation involves not only the creation of something new but also the process of developing and testing it, finding a market for it, and ultimately producing and delivering it to customers. In summary, invention is the idea or creation of something new, while innovation is the process of making that idea a reality and bringing it to market.

What Are Myths About Innovation?

Innovation only happens in Silicon Valley

The idea that innovation only happens in certain geographic

locations, such as Silicon Valley, is a myth. Innovation can happen anywhere, and many successful innovations have come from small towns, rural areas, and emerging markets.

Innovation is only for tech companies

Innovation is not limited to technology companies, it can happen in any industry and organization. Many non-tech companies, such as Procter & Gamble, have long histories of successful innovation.

Innovation requires a big budget

Innovation can happen on any budget. Many successful innovations have been developed on a shoestring budget, using open-source software, and other low-cost tools.

Innovation only comes from individuals

Innovation often comes from teams, not just individuals. Collaboration and diversity of ideas can lead to more successful innovations.

Innovation is only about creating new products

Innovation can also be about improving existing products,

services, and processes. Incremental innovation can lead to significant improvements and cost savings over time.

Innovation only happens in R&D department

Innovation can happen anywhere in the organization, not just in the R&D department. Encouraging a culture of innovation and involving employees from all levels and departments can lead to more successful innovations.

Innovation is a one-time event

Innovation is a continuous process. It requires ongoing effort to identify new opportunities, generate ideas, and implement new solutions.

What Are Different Types Of Innovation?

Product innovation and service innovation refers to the introduction of new or improved products, services, or goods, while process innovation refers to the introduction of new or improved methods of production, delivery or distribution.

Product Innovation

Product innovation can be a new product or feature

that addresses customer needs in a new way, or an existing product that has been improved to offer better performance, quality or value. Examples of product innovation can be a new mobile phone, a new car model, or a new type of food.

Service Innovation

The development and implementation of new or improved services that create value for customers. This can include new service models, such as subscription-based services or self-service options, as well as improvements in the delivery of existing services, such as online service portals or mobile apps.

Process Innovation

Process innovation refers to improvements made to the way in which goods or services are produced, delivered or distributed. This can include improvements in manufacturing processes, logistics, supply chain management or customer service.

Process innovation can be incremental, disruptive or radical, and it can lead to cost savings, increased efficiency, improved quality or faster delivery times. Examples of process innovation can be a new manufacturing process, a new distribution channel, or a new way of providing customer service.

Product innovation and service innovation can help companies to differentiate themselves from their competitors and attract new customers, while process innovation can help companies to improve efficiency and reduce costs. They are important for companies to stay competitive and drive growth. Companies can also leverage both product and process innovation together to achieve their goals.

What Are Different Approaches To Innovation?

Innovation can take many forms, from small improvements to existing products or processes, to entirely new technologies or business models. Understanding the different approaches to innovation can help to identify opportunities for growth and stay competitive in their industries. There are several types of innovation, including incremental, disruptive, and radical innovation, as well as open and social innovation, and business model innovation.

Incremental Innovation

Incremental innovation involves making small improvements to existing products, services, or processes. These improvements can be in the form of new features, better design, or cost reductions.

For instance, Electric cars have been around for a while, but their batteries were not able to hold a charge for very long, which limited their range. Over time, through incremental innovation, battery manufacturers have been able to improve the energy density of the batteries, which means they can store more energy in a smaller space. This has led to a significant increase in the range of electric cars, making them more practical for everyday use.

This type of innovation can be seen in the example of Tesla, which has been working on improving the range of its electric cars for years. Each new model of car has had a longer range than the previous one, and the company has also been working on developing new battery technologies, such as lithium-ion batteries, to further improve the range of its cars. This incremental innovation has allowed Tesla to stay competitive in the electric car market and has helped to make electric cars more viable for consumers.

Disruptive Innovation

Disruptive innovation involves creating new technologies or business models that fundamentally change the way an industry operates. Disruptive innovations often start in niche markets and then disrupt established markets.

For instance, Uber and Lyft introduced a new business model of ride-sharing services that disrupted the traditional taxi and car rental industries. Instead of hailing a taxi on the street or calling a car rental company, customers can use a smartphone app to request a ride and track the location of the driver in real-time.

The disruptive innovation of Uber and Lyft was the use of technology to create a new way of providing transportation services, by connecting riders with drivers through a mobile app and making the process more convenient, efficient and affordable than traditional taxis. It also created a new market of ride-sharing services, where the company itself doesn't own any cars, but instead uses independent contractors as drivers, which allows them to operate with a lower cost structure. This innovation has led to a significant shift in the transportation industry and has challenged traditional taxi and car rental companies to adapt to the new reality.

Radical Innovation

Radical innovation approach involves creating something entirely new, such as a new product, service, or process that does not currently exist. Radical innovations can lead to the creation of new markets and industries.

For instance, T20 is a shortened version of the traditional cricket format, which is played for 20 overs per side. The introduction of T20 cricket in 2003 marked a radical change in the way the sport was played and watched.

Traditionally, cricket was known for its long and drawn-out matches, with five-day test matches being the norm. T20 cricket, on the other hand, is fast-paced and action-packed, with a match lasting around three hours. This new format is designed to attract new fans, especially younger generations, who may have been put off by the slow pace of traditional cricket.

T20 cricket has had a significant impact on the sport, it has attracted new fans, especially in countries where cricket wasn't popular before, and it has also created new leagues and tournaments, such as the Indian Premier League and Big Bash League, which have become major events on the cricket calendar. Additionally, T20 cricket has also led to changes in the way the game is played, with players now focusing more on power-hitting and aggressive play. This change in the format has led to the creation of new opportunities for players and teams, and it has also led to a change in the way the sport is marketed and promoted.

Open innovation

Open innovation approach involves engaging with external partners, such as customers, suppliers, or other companies, to generate and develop new ideas. Nowadays, governments are using crowdsourcing platforms to generate new ideas and solve problems. It allows them to tap into the collective knowledge and expertise of many people, often including citizens, businesses, and other organizations, to generate new ideas and solve problems.

In 2013, Netherlands launched the platform "Crowdsourcing the Netherlands". The platform allows citizens to submit ideas for new policy initiatives, and then vote on the best ideas. The government then uses the top-voted ideas to inform their policy-making decisions. Another example is the UK's "GovTech Catalyst" launched in 2016, which is a platform that allows public sector organizations to collaborate with technology companies and entrepreneurs to develop new digital solutions to public services. The platform is designed to accelerate the delivery of public services and create a more efficient, effective and transparent government.

By using open innovation techniques, governments can access a wider pool of ideas and perspectives, and find new and better ways to solve problems and improve services. This not only can lead to more effective and efficient government, but also can increase citizens engagement and trust in the government.

Other Approaches

It's worth noting that different approaches of innovation can overlap and an innovation can be a combination of these types. The approach to innovation a company or organization chooses to pursue often depends on its goals, resources, and industry. Few other approaches of innovation, include:

Social innovation: Social innovation approach refers to the introduction of new ideas or solutions to address societal challenges, such as poverty, inequality, and environmental issues.

An example of social innovation in the dairy industry in India is the establishment of the "Amul" model of dairy co-operatives. Amul, which stands for Anand Milk Union Limited, is a dairy cooperative in the Indian state of Gujarat. The co-operative was established in 1946 and is now one of the largest dairy cooperatives in the world, with more than 3.6 million milk producers.

The Amul model is based on the principles of collective action and self-help, where milk producers are organized into village-level cooperatives, which are then affiliated to a larger district-level federation. The cooperatives are owned and controlled by the milk producers, who collectively decide on the management and marketing of their milk.

The Amul model has had a significant impact on the dairy industry in India, and has been credited with transforming the lives of millions of small-scale milk producers. By providing them with a stable source of income and access to markets, the cooperatives have helped to improve their livelihoods and raise their standard of living. The Amul model has also helped to improve the quality and quantity of milk produced in India and has made the country one of the world's largest milk producers.

The Amul model has been recognized as a social innovation because it also has been able to empower women, as they are often the primary milk producers in rural India, and by providing them with a stable source of income, it has helped to improve their status and position in society.

Business Model Innovation: Business model innovation approach refers to the creation of new ways of doing business, such as new revenue streams, distribution channels, or partnerships. This can include changes to pricing models, the way products or services are delivered, or the way in which customers are engaged.

For instance, streaming services like Netflix and Spotify allow customers to access a wide variety of movies, TV shows, and music on-demand, and pay a monthly subscription fee rather than buying or renting individual titles. This business model innovation has disrupted the traditional entertainment industry, which relied on the sale of physical media and advertising revenue.

Netflix, which started as a mail-order DVD rental service, but later pivoted to streaming service in 2007, has been a pioneer in this field, by creating its own exclusive content, with shows like Stranger Things and The Crown, and by investing in producing more and more original content. This not only has allowed them to attract and retain customers, but also has created new revenue streams for the company.

Spotify, which launched in 2008 as a music streaming service, has also disrupted the music industry by making it easy for customers to access a vast library of music for a monthly subscription fee, and it also has created new revenue streams for the company through its premium subscription service, which offers ad-free listening and offline playback.

Netflix and Spotify show how business model innovation can disrupt traditional industries and create opportunities for growth and revenue. Streaming services have changed the way people consume and pay for entertainment and have opened new market opportunities and created new revenue streams for companies in the industry.

Marketing Innovation: Marketing innovation approach involve the development of new marketing strategies and techniques to reach and engage customers. It includes new forms of advertising, such as influencer marketing or virtual reality, or new ways of segmenting and targeting customers.

One example of marketing innovation is the use of Influencer Marketing. It is a strategy in which brands partner with individuals who have a large following on social media platforms to promote their products or services. The influencers use their personal social media accounts to post about the brand, share their experiences, and provide their followers with a personal recommendation.

The use of influencer marketing as a form of marketing innovation allows companies to reach a specific target audience, increase brand awareness and credibility, and provide a more authentic form of advertising, as the influencers are seen as a trusted source of information by their followers. Additionally, influencer marketing can also help companies to reach new and younger audiences, as many social media users are millennial and Generation Z.

Fashion brand, Fashion Nova, used influencer marketing and partnered with various celebrities, especially from the entertainment industry, to promote their products on social media platforms like Instagram, TikTok, Twitter, etc. This approach has helped them to increase their brand awareness and credibility, as well as drive sales and revenue. Each innovation approach has its own characteristics and can be used to achieve different goals depending on the context. Companies can also use a combination of these types of innovation to achieve their objectives and stay competitive.

What Are The Main Stages Of Innovation Lifecycle?

The innovation process is a structured approach to identifying and developing new ideas, and turning them into successful products, services, or processes. The process typically includes several stages, such as idea generation, feasibility analysis, prototyping, testing, and implementation. It also involves input and collaboration from different departments and stakeholders, such as R&D, marketing, engineering, and management.

The innovation process is an ongoing and iterative one, and it requires a combination of creativity, critical thinking, and a willingness to take risks. It is an essential tool for organizations to stay relevant and adapt to the changing environment, as well as to create new value for customers, shareholders and society.

The innovation process typically involves the following steps:

Idea Generation

This is the first step in the innovation process, where ideas for new products, services, or processes are generated. It is an important step in the innovation process, as it involves identifying and developing new and creative ideas that can be used to solve problems or create new opportunities. This can be done through brainstorming sessions, customer feedback, market research, or other methods.

There are several methods and techniques that can be used for idea generation, including:

Brainstorming: This is a group activity where a team of people come together to generate as many ideas as possible in a short amount of time. Brainstorming sessions can be structured or unstructured and can be done in person or remotely.

Mind mapping: This is a visual tool that helps to organize and connect ideas. It involves creating a diagram or map that shows the relationship between different ideas and concepts. Mind mapping can be done individually or in a group.

SCAMPER: This is a tool that stands for Substitute, Combine, Adapt, Modify, Put to another use, Eliminate, Reverse. It helps to generate ideas by asking questions about how a product, service or process can be improved by applying these seven different techniques.

Reverse engineering: This involves taking something that already exists and figuring out how to improve it or use it in a different way.

Customer feedback: This involves gathering feedback from customers to understand their needs, wants and pain points. This feedback can be used to generate ideas for new products, services, or process improvements.

Crowdsourcing: This involves inviting many people to submit ideas, suggestions or solutions to a specific problem or opportunity. Crowdsourcing can be done online or offline, and the ideas can be evaluated and filtered through a voting process or an expert panel.

These are just a few examples of methods that can be used for idea generation in the innovation process. The best approach will depend on the specific problem or opportunity, the resources available and the goals of the organization.

Feasibility Analysis

Once ideas have been generated, they are evaluated to determine their feasibility. This includes assessing the potential market size, competition, and costs associated with developing and launching the idea.Feasibility analysis is an important step in the innovation process, as it involves evaluating the practicality and potential success of an idea or solution. Feasibility analysis can help to identify potential problems or obstacles that need to be addressed before proceeding with the implementation of the idea.

There are several factors that are typically considered during a feasibility analysis, including:

Technical feasibility: This involves evaluating whether the necessary technology and resources are available to implement the idea.

Economic feasibility: This involves evaluating whether the idea is financially viable, and whether the expected benefits will outweigh the costs.

Operational feasibility: This involves evaluating whether the idea can be integrated into the existing operations and processes of the organization.

Legal and regulatory feasibility: This involves evaluating whether the idea is compliant with relevant laws and regulations.

Environmental and social feasibility: This involves evaluating the potential impact of the idea on the environment and society.

An example of feasibility analysis in the innovation process is the development of a new product. The product development team would conduct a feasibility analysis to evaluate the technical feasibility of the product by determining if the necessary technology and resources are available to produce the product, the economic feasibility by evaluating the potential revenue and costs of the product, the operational feasibility by determining if the product can be produced in the company's existing facilities, the legal and regulatory feasibility by ensuring the product meets all necessary regulations, and the environmental and social feasibility by evaluating the impact the product will have on the environment and society.

Feasibility analysis can help to identify potential problems or obstacles early on in the innovation process and can help to ensure that resources are not wasted on ideas that are unlikely to succeed. It helps to minimize the risk and maximizes the chances of success of the innovation.

Prototyping

If an idea is determined to be feasible, a prototype is developed. This allows the idea to be refined before moving to the next stage of the process. Prototyping allows for the testing and evaluation of an idea or solution in a tangible form. A prototype is a preliminary version of a product, service or process that allows for the testing of its feasibility, function, and design.

There are several types of prototypes that can be used in the innovation process, including:

Conceptual prototypes: These are simple and inexpensive mock-ups that are used to test the basic concept of an idea. They can be made with materials such as paper, cardboard, or clay and are usually used to test the usability and design of a product.

Functional prototypes: These are more detailed and complex mock-ups that are used to test the functionality of a product or service. They can be made with materials such as plastic or metal and are usually used to test the technical feasibility and performance of a product or service.

User experience prototypes: These are interactive mock-ups that are used to test the user experience of a product or service. They can be made with software such as wireframe tools, and are usually used to test the usability, design and user interaction of a product or service.

An example of prototyping in the innovation process is the

development of a new mobile application. The development team would create a conceptual prototype using wireframe tools to test the basic layout and design of the application. They would then create a functional prototype using the actual programming languages and test the application's functionality and performance. Finally, they would create a user experience prototype using interactive wireframes and test the user's experience and interaction with the application.

Prototyping allows for the testing and evaluation of an idea in a tangible form, and it helps to identify any potential problems or obstacles early in the innovation process. It also enables the team to get feedback from customers and stakeholders and adjust before the final version is developed.

Testing And Validation

The prototype is tested and validated to ensure that it meets the needs of customers and the market. Feedback is collected and used to further improve the product, service, or process.

There are several types of testing and validation that can be used in the innovation process, including:

Alpha testing: This is an internal testing stage conducted by the development team. Alpha testing is done to identify and fix any bugs or errors in the product, service, or process.

Beta testing: This is an external testing stage that is conducted by a small group of customers or users. Beta testing is done to gather feedback on the product, service or process and identify any issues or areas of improvement.

User acceptance testing: This is a testing stage that is conducted by the end-users of a product, service, or process. User acceptance testing is done to ensure that the final product meets the requirements and expectations of the end-users.

Performance testing: This is a testing stage that is conducted to evaluate the performance of a product, service or process under different conditions and loads.

Safety testing: This is a testing stage that is conducted to ensure that a product, service, or process meets the safety standards and requirements.

An example of testing and validation in the innovation process is the development of a new medical device. The development team would conduct alpha testing to ensure that the device meets the technical requirements and has no bugs or errors. They would then conduct beta testing with a small group of hospitals and gather feedback on the device's usability and performance. Next, they would conduct user acceptance testing with the end-users, the medical staff, to ensure that the device meets their requirements and expectations.

Finally, they would conduct performance and safety testing to ensure that the device works as intended under different conditions and loads and meets the safety standards and requirements.

Testing and validation allow for the evaluation of the performance, reliability and safety of a product, service, or process. This step is crucial to ensure that the final product meets the requirements and expectations of the customers, stakeholders, and the organization itself. This step also allows for the identification and resolution of any issues or bugs before the final version is released.

Commercialization

If the idea is determined to be viable, it is commercialized and launched into the market. This includes developing a business plan, marketing strategy, and production plan.

Commercialization involves the process of introducing a new product, service or process to the market and making it available to customers. This step includes all the activities necessary to bring the product to the market such as market research, product development, marketing, sales and distribution, pricing, and after-sales service.

There are many methods of commercialization, including:

Licensing: This involves granting the rights to use a product, service, or process to another company. This can be a good option for companies with limited resources or for technologies that require significant investment to bring to market.

Joint venture: This involves forming a partnership with another company to share the risk and costs of commercialization.

Spin-off: This involves creating a new company to commercialize the product, service, or process.

Direct selling: This involves selling the product, service or process directly to customers.

An example of commercialization in the innovation process is the development of a new renewable energy technology. The development team would conduct market research to determine the size and characteristics of the target market and the potential demand for the technology.

They would then develop a marketing and sales strategy, which could include a combination of direct selling and licensing to other companies. They would also determine the pricing strategy and create a distribution network to make the technology available to customers. Finally, they would develop an after-sales service plan to provide support and maintenance to customers.

Commercialization allows the product, service or process to reach the market and generate revenue, and it also allows the organization to recoup the investment made in the innovation process. Additionally, it also allows the organization to measure the success of the innovation by looking at the market acceptance, financial performance and customer feedback.

Continual Improvement

The innovation process doesn't end with commercialization, it's an ongoing process. After launch, the product or service is monitored, and improvements are made as needed.

Continual improvement is an ongoing process in the innovation process, as it involves the process of monitoring and evaluating the performance of a product, service or process and making changes to improve it over time. This step is crucial to ensure that the product, service or process meets the evolving needs of customers, stakeholders, and the market.

There are several methods of continual improvement, including:

Benchmarking: This involves comparing the performance of a product, service, or process with that of other similar products, services or processes in the market. This allows for the identification of areas for improvement and the implementation of best practices.

Kaizen: This is a Japanese word meaning "improvement" or "change for the better", which is a philosophy focused on continuous improvement and employee involvement.

Total Quality Management (TQM): This is an approach to management that aims to achieve customer satisfaction and continuous improvement by involving all employees in the quality improvement process.

An example of continual improvement in the innovation process is the development of a new software application. The development team would use benchmarking to compare the performance of the application with similar software in the market, identify areas for improvement and implement best practices. Additionally, they would conduct regular user testing and gather feedback from customers to evaluate the application's performance and make changes based on their feedback. Lastly, they would use TQM approach to involve all employees in the quality improvement process and ensure that the application meets the evolving needs of customers and the market.

Continual improvement allows the product, service, or process to meet the evolving needs of customers, stakeholders, and the market, and it also allows the organization to adapt to the changing environment and stay competitive.

It allows the organization to measure the performance of the innovation over time and adjust as necessary. It's worth noting that this process is not linear, and it may involve going back and forth between steps or skipping certain steps depending on the specific situation. Additionally, in some cases, the process of innovation may be more incremental, with small improvements made to existing products or services, rather than creating something entirely new.

2

URBAN INNOVATIONS FOR CITY GOVERNMENTS

Urban innovation is not about creating new things, it's about creating new ways of doing things. It's about rethinking the way we live, work and play in our cities, and making them more sustainable, livable and equitable.

- Jan Gehl

Urban innovation is like finding new and better ways to make the cities we live in more fun, safe, and comfortable for everyone. For example, imagine you live in a city, and you like to ride your bike, but there aren't many bike lanes, so it's not always safe for you to ride. Urban innovation can help make more bike lanes for you and other kids to ride safely. Or maybe there's a park near your house, but it's not very pretty or fun to play in. Urban innovation can help make the park more fun and beautiful for you and your friends to play in. Urban innovation is like making the city a big playground for everyone to enjoy.

Defining Urban Innovation

Urban innovation refers to the introduction of new and improved technologies, products, services, policies, or practices that address the challenges and opportunities of urban environments. Urban innovation can take many forms and can be used to improve the livability, sustainability, and resilience of cities.

Urban innovation can include:

Citizen Innovation

Such as community-based initiatives and citizen engagement programs, which address social and economic challenges in urban areas.

Digital Innovation

The use of digital technologies, such as artificial intelligence, big data, and the Internet of Things, to improve products, services, and business processes. Smart city technologies such as smart grid, smart transportation, and smart buildings, which use sensors and data analytics to improve the efficiency and sustainability of urban systems.

Organizational Innovation

The introduction of new ways of organizing and managing an organization, such as new forms of leadership, new organizational structures, or new ways of motivating and engaging employees.

Regulatory and Policy Innovation

The development and implementation of new regulations and policies to promote sustainable development and improve the livability of cities or countries. Such as innovative financing mechanisms, and new regulations or policies that promote sustainable development and improve the livability of cities.

Sustainability Innovation

Such as green roofs, electric vehicles, energy efficient lighting, which improve environmental performance of cities.

Urban Planning and Design Innovation

Such as bike lanes and pedestrian-friendly streets, which improve the livability of cities.

Urban innovation can help to improve the quality of life for urban residents, by addressing issues such as housing, transportation, public spaces, energy, and water. It can also help to create more sustainable and resilient urban environments, by reducing greenhouse gas emissions, improving energy and water efficiency, and promoting sustainable mobility and land use patterns.

Focus Areas For Urban Innovations

As urban areas continue to grow and evolve, cities face a wide range of complex and interrelated challenges, including climate change, air and water pollution, social inequality, transportation, and economic development. To address these challenges, cities are increasingly looking to innovate and develop new solutions that can improve the quality of life for residents, and create new opportunities for growth and development.

There are a number of focus areas for urban innovation that are currently being explored by cities around the world. These include:

Citizen Innovation

Citizen innovation refers to the active involvement of citizens in the co-creation, co-design and co-delivery of public services, policies and projects that address urban challenges.

This type of innovation is based on the principle of open and inclusive governance, where citizens are not only recipients of services, but also active participants in the design and delivery of these services.

It is based on the idea that citizens, with their unique perspectives have valuable knowledge, skills, and experience that can contribute valuable insights and ideas to improve public services and to develop new and creative solutions to the complex and interrelated challenges faced by cities. By involving citizens in the innovation process, cities can increase transparency, trust, and inclusivity, and achieve more equitable and sustainable solutions.

There are several methods of citizen innovation in cities, including:

Citizen engagement and participation: This method involves the engagement of citizens, through activities such as public consultations, and workshops to provide feedback on policy proposals or service delivery and involve citizens in decision-making processes.

Citizen co-creation: This method involves the active involvement of citizens in the design and delivery of services and policies, working alongside government officials and experts.

Crowdsourcing and open innovation: This method involves the use of digital platforms to gather ideas from a wide range of citizens and stakeholders to solve urban problems and to find solutions.

Living labs and urban experimentation: This method involves the use of real-world environments to test and evaluate new solutions, with the active involvement of citizens and stakeholders.

Social innovation and social entrepreneurship: This method involves the development of new solutions by citizens and civil society organizations, to address social and environmental challenges.

Citizen innovation can help cities to achieve more equitable and sustainable solutions, by involving citizens in the innovation process and taking into account their perspectives, needs and aspirations. By involving citizens in the innovation process, governments can gain valuable insights and perspectives, and create more responsive, effective and sustainable services and policies. Additionally, by involving citizens in the innovation process, cities can also increase transparency, trust and inclusivity, and ensure that the solutions developed are aligned with the needs and expectations of the community.

Digital Innovation

Digital innovations are technologies and solutions that leverage digital technologies such as such as artificial intelligence, big data, and the Internet of Things, to improve the efficiency, effectiveness, and sustainability of urban services and infrastructure. These innovations can help cities to address a wide range of urban challenges in transportation, energy, water management, and public safety.

Examples of digital innovation include:

Artificial intelligence: the use of algorithms and machine learning to automate tasks, improve decision-making, and create new products and services.

Big data: the use of large data sets and analytics to gain insights, improve decision-making, and create new products and services.

The Internet of Things (IoT): the use of connected devices and sensors to collect and share data, automate tasks, and create new products and services.

Cloud computing: the use of remote servers and data storage to access and share data and resources, improve collaboration and reduce costs.

Blockchain: the use of decentralized digital ledger technology to securely record and share data, automate tasks and create new products and services

Virtual and augmented reality: the use of computer-generated images and other sensory input to create immersive experiences, and create new products and services.

Some use cases for digital innovations in cities include:

Smart traffic management systems: these use sensors and cameras to collect data on traffic flow and patterns and optimizes traffic signals and reduce congestion.

Smart lighting systems: these use sensors and cameras to monitor and adjust the brightness of streetlights, reducing energy consumption and costs.

Smart waste management systems: these use sensors and cameras to monitor the fill level of waste receptacles and optimize collection routes to reduce waste and costs.

Smart parking systems: these use sensors and cameras to monitor the availability of parking spaces and provide real-time information to drivers to reduce congestion and emissions.

Smart buildings: these use sensors, cameras, and building management systems to monitor and optimize the energy consumption of buildings and improve the comfort and safety of occupants.

Digital innovations can help cities to improve the efficiency and effectiveness of urban services, reduce costs and emissions, and create new opportunities for economic growth and development. Additionally, by leveraging the capabilities of digital technologies, cities can create more responsive, transparent, and citizen-centered services.

Organizational Innovation

Organizational innovation can help city governments to become more efficient, responsive and effective in delivering services to citizens, and addressing the challenges of urbanization.

There are several examples of organizational innovation in city governments:

Lean government: This is based on the principles of Lean manufacturing and focuses on reducing waste, improving efficiency and increasing quality in government operations. It involves using process mapping and other tools to identify inefficiencies and opportunities for improvement, and then implementing changes to improve the performance of government services.

Agile government: This approach is based on the principles of Agile software development and focuses on flexibility, speed and adaptability in government operations. It involves breaking down complex projects into smaller, more manageable tasks, and continuously testing and adjusting to improve performance.

Results-oriented management: This approach focuses on setting clear goals and objectives, measuring performance and holding managers accountable for achieving results. It involves using data and performance metrics to track progress and make adjustments to improve the effectiveness of government services.

Collaborative governance: This approach focuses on

involving citizens and other stakeholders in the decision-making process and co-creating solutions to address urban challenges. It involves creating platforms for citizen engagement and collaboration, such as online portals, public meetings, or citizen advisory boards.

By implementing organizational innovation, city governments can improve the quality of services they provide, increase citizen satisfaction, and provide better responses to the challenges that the cities face. By involving citizens and stakeholders, they can also increase trust and transparency and make the process of governance more inclusive.

Regulatory and Policy Innovation

Regulatory and policy innovation refers to the development and implementation of new regulations and policies that promote sustainable development and improve the livability of cities. These innovations can include new zoning laws, building codes, and transportation policies that encourage sustainable development, or new financial incentives or regulatory frameworks that promote sustainable practices.

Examples of regulatory and policy innovation in cities include:

Green building codes: regulations that set standards for energy efficiency, water conservation, and other environmental performance in new construction and renovations.

Smart growth policies: policies that promote compact, mixed-use development and discourage sprawl, to reduce vehicle use and promote more sustainable land use patterns.

Sustainable transportation policies: policies that promote the use of public transit, biking, and walking, and discourage the use of cars, to reduce greenhouse gas emissions and improve air quality.

Carbon pricing: policies that put a price on carbon emissions, discourage the use of fossil fuels and promote the use of clean energy.

Sustainable procurement policies: regulations that

require the government to purchase goods and services that meet certain environmental or social standards.

Adaptive reuse policies: regulations that encourage the revitalization and reuse of existing buildings and infrastructure, which can lead to more sustainable and livable urban spaces.

Regulatory and policy innovation can help cities to address the challenges of urbanization, such as climate change, air pollution, and inequality. By implementing new regulations and policies, cities can create an enabling environment for sustainable development and create new opportunities for economic growth and development. Additionally, by involving citizens and stakeholders in the policy-making process, cities can increase transparency, trust, and inclusivity.

Sustainability Innovation

It refers to the development and implementation of new and creative solutions that improve the environmental, social, and economic sustainability of urban areas. These innovations can range from new technologies and infrastructure to policies and regulations and can address a wide range of urban challenges such as climate change, air and water pollution, and social inequality.

Examples of sustainability innovation in cities include:

Smart city technologies: these include technologies such as IoT, big data, and cloud computing that are used to improve the efficiency and sustainability of urban services such as transportation, energy and waste management.

Green infrastructure: this includes the incorporation of natural systems such as green roofs, urban forests and rain gardens into the built environment to improve air and water quality, reduce heat island effects and promote biodiversity.

Sustainable transportation: this includes the development of low-carbon transportation options such as electric buses, bike-sharing and ride-hailing services, as well as the implementation of policies to promote active transportation such as walking and cycling.

Energy-efficient buildings: this includes the design and construction of buildings that use less energy and water, and generate less waste and greenhouse gas emissions.

Circular economy: this includes the development of systems and policies that promote the use of resources in a more efficient and sustainable way, such as by encouraging the recycling and re-use of materials, and the development of closed-loop systems.

Sustainability innovation can help cities to address the challenges of urbanization and improve the livability of urban areas. By implementing new technologies, policies and regulations, cities can create an enabling environment for sustainable development, and create new opportunities for economic growth and development. Additionally, by involving citizens and stakeholders in the innovation process, cities can increase transparency, trust and inclusivity.

Urban Planning and Design Innovation

Urban planning and design innovation refers to the development and implementation of new or improved urban design concepts, techniques, technologies, and policies that address the challenges and opportunities of urban environments. Urban planning and design innovation can be used to improve the livability, sustainability, and resilience of cities and create more livable and sustainable urban spaces.

Urban planning and design innovation can take many forms, such as:

Smart city planning: the use of data analytics, simulation models, and other digital tools to inform urban planning decisions and design more efficient, sustainable, and livable cities.

Sustainable urban design: the use of green infrastructure, low-impact development, and sustainable transportation design to create more livable, resilient, and sustainable urban spaces.

Community-based planning and design: the use of participatory planning and design methods to engage local communities and stakeholders in the planning and design process, which can lead to more livable, sustainable, and equitable urban spaces.

Adaptive reuse: the revitalization and reuse of existing buildings and infrastructure, which can lead to more sustainable and livable urban spaces.

Urban planning and design innovation is important because it can help to address the challenges of urbanization and create more livable, sustainable, and resilient urban spaces. By using innovative planning and design concepts, techniques, technologies, and policies, cities can reduce their environmental impact, improve the livability and resilience of urban spaces, and create new opportunities for growth and development.

Key Stakeholders In Urban Innovations

Key stakeholders in urban innovation can vary depending on the specific innovation and context, but some common groups include:

City officials and staff

City officials and staff are key stakeholders, as they are

responsible for the implementation and management of urban innovations.

Residents and community groups

Residents and community groups are key stakeholders, as they are directly impacted by urban innovations and can provide valuable feedback and perspectives.

Private sector

Private sector organizations and businesses are key stakeholders, as they can provide funding, expertise and support for urban innovations.

Non-profit organizations

Non-profit organizations and advocacy groups are key stakeholders, as they can provide expertise and support on specific issues related to urban innovation.

Academic and research institutions

Academic and research institutions are key stakeholders, as they can provide research and expertise on urban innovation and related topics.

Environmental groups

Environmental groups are key stakeholders, as they can provide expertise on the environmental aspects of urban innovation and how it relates to sustainability.

Government agencies

Government agencies are key stakeholders, as they can provide funding, regulations and support for urban innovations.

Financial Institutions and investors

Financial institutions and investors are key stakeholders, as they can provide funding and support for urban innovation projects, particularly for infrastructure and other large-scale projects.

By engaging with these key stakeholders, cities can ensure that new ideas are inclusive, responsive and beneficial to the community, and that the solutions developed are sustainable over the long term.

Managing Urban Innovations

Urban innovation management is the process of planning, developing, implementing, and evaluating new ideas, policies, and technologies that can improve the livability, sustainability, and resilience of cities. This process often involves multiple stakeholders, including city governments, businesses, community organizations, and citizens, working together to identify problems and opportunities, generate new ideas, and implement solutions. Urban innovation management can involve a wide range of activities, such as:

Problem Identification and Opportunity Identification

Identifying problems and opportunities in managing cities is the first step in the urban innovation management process. Here are a few ways to approach problem and opportunity identification in managing cities:

Data analysis: Analyze data on a variety of urban issues, such as population growth, demographics, economic activity, infrastructure, and environmental conditions, to identify areas where problems and opportunities may exist.

Stakeholder engagement: Engage with a wide range of stakeholders, such as city governments, businesses, community organizations, and citizens, to gain a better understanding of the problems and opportunities they see in the city.

Benchmarking: Compare the city's performance on various urban issues to other cities, to identify areas where the city is doing well and areas where it could improve.

Scenario planning: Use scenario planning to explore different future scenarios and identify potential problems and opportunities that may arise as a result of changing population, economic, environmental, or technological conditions.

Crowdsourcing: Use crowdsourcing to gather ideas and feedback from citizens, businesses, and other stakeholders on urban issues, and identify the most pressing problems and opportunities in the city.

Innovation contests: Organize innovation contests to

encourage citizens, students, and businesses to submit ideas and solutions for addressing urban problems and opportunities.

By using these methods, cities can gain a better understanding of the problems and opportunities they face, and can better identify areas where urban innovation can make a meaningful impact. Furthermore, involving a wide range of stakeholders in the process of problem and opportunity identification can help to ensure that the solutions developed are inclusive and responsive to the needs of the community.

Idea Generation

Generating ideas is the next step in the urban innovation management process. Here are a few ways to approach idea generation in urban innovation management:

Brainstorming: Hold brainstorming sessions with a diverse group of stakeholders, such as city officials, community leaders, business owners, and citizens, to generate a wide range of ideas on how to address urban problems and opportunities.

Hackathons: Hold hackathons or design sprints to bring together a diverse group of people with different skills and expertise to generate and develop ideas on how to address urban problems and opportunities.

Open innovation: Encourage external organizations and individuals to submit ideas for addressing urban problems and opportunities through open innovation platforms or contests.

Innovation workshops: Hold workshops or boot camps to provide training and resources to citizens, businesses, and other stakeholders to help them generate and develop ideas on how to address urban problems and opportunities.

Research and development: Invest in research and development to identify new technologies, policies, and practices that can be used to address urban problems.

Learning visits: Organize learning visits to other cities to study and learn from other cities' innovative solutions and best practices.

Case studies: Develop case studies on other cities to understand the problems they've solved, what measures they used and how they implemented them.

Citizen engagement: Involve citizens in the idea generation process by holding public meetings, town hall meetings, and using social media to gather feedback and ideas on how to address urban problems and opportunities.

By using these methods, cities can generate a wide range of ideas from a diverse group of stakeholders and increase the chances of identifying innovative solutions that can address urban problems and opportunities. Furthermore, by involving citizens in the idea generation process, cities can ensure that the solutions developed are inclusive and responsive to the needs of the community.

Feasibility Analysis

Feasibility analysis helps cities to evaluate whether new ideas are viable and practical to implement. Here are a few ways to approach feasibility analysis in urban innovation management:

Technical feasibility: Assess the technical feasibility of new ideas by evaluating whether the necessary technology and infrastructure are available to implement the idea, and whether the idea is compatible with existing systems and processes.

Financial feasibility: Assess the financial feasibility of new ideas by evaluating the costs and benefits of the idea, and determining whether the idea is financially viable over the short and long term.

Social feasibility: Assess the social feasibility of new ideas by evaluating how the idea will impact different groups of people, and whether the idea is acceptable and desirable to the community.

Environmental feasibility: Assess the environmental feasibility of new ideas by evaluating the potential impact of the idea on the environment and whether the idea is sustainable and resilient.

Legal and regulatory feasibility: Assess the legal and regulatory feasibility of new ideas by evaluating whether the idea is in compliance with existing laws, regulations and policies.

Cost-benefit analysis: Do a cost-benefit analysis to evaluate the costs and benefits of the idea, including both monetary and non-monetary costs and benefits, such as environmental and social impacts.

Pilot testing: Run pilot tests of new ideas to gather data on the feasibility of the idea in real-world conditions.

By using these methods, cities can gain a better understanding of the feasibility of new ideas and increase the chances of identifying solutions that are viable and practical to implement. Furthermore, by conducting feasibility analysis that considers technical, financial, social, environmental, legal and regulatory, cities can ensure that the solutions developed are inclusive, sustainable and resilient.

Prototyping

Prototyping allows cities to test and refine new ideas in

real-world settings before implementing them on a larger scale. Here are a few ways to approach prototyping in urban innovation management:

Rapid prototyping: Use low-cost and quick methods to create a physical or digital prototype of the new idea, to test its basic functionality and usability.

Pilot testing: Test the new idea in a small-scale pilot project, to gather data on its feasibility and effectiveness in real-world conditions.

Simulation: Create computer simulations of the new idea to test its performance in different scenarios and to evaluate its potential impact.

Scale models: Build scale models of the new idea to test its physical characteristics, such as size, shape and layout.

Co-creation: Involve end-users, stakeholders, and other experts in the prototyping process to gather feedback and improve the design of the prototype.

Iterative prototyping: Refine the prototype through an iterative process, by incorporating feedback and making adjustments as needed.

Living lab: Test the prototype in a real-world setting using a Living Lab approach, which enables stakeholders to test and evaluate the prototype in real-world conditions.

By using these methods, cities can test and refine new ideas in real-world settings before implementing them on a larger scale, which can increase the chances of success and reduce the risks of implementation. Furthermore, by involving stakeholders and end-users in the prototyping process, cities can ensure that the solutions developed are responsive to the needs of the community and more likely to be adopted.

Testing and Validation

Testing and validation allows cities to gather data on the effectiveness and impact of new ideas before implementing them on a larger scale. Here are a few ways to approach testing and validation in urban innovation management:

Pilot testing: Test the new idea in a small-scale pilot project, to gather data on its feasibility, effectiveness, and impact in real-world conditions.

Field testing: Test the new idea in a real-world setting, to gather data on its performance, durability and impact in real-world conditions.

Controlled experiments: Use controlled experiments to test the new idea under different conditions, to gather data on its effectiveness and impact.

Surveys and interviews: Use surveys and interviews to gather feedback and data on the new idea from stakeholders and end-users.

User testing: Gather feedback and data on the new idea from users, to evaluate its usability and effectiveness.

Data analysis: Collect and analyze data on the new idea, to evaluate its effectiveness and impact.

Impact assessment: Conduct an impact assessment of the new idea, to evaluate its long-term effects on the city, its inhabitants and the environment.

These methods help cities gather data on the effectiveness and impact of new ideas before implementing them on a larger scale, which can increase the chances of success and reduce risks of implementation. By conducting testing and validation that considers the perspectives of different stakeholders and end-users, cities can ensure that the solutions developed are inclusive, responsive and beneficial to the community.

Implementation

Implementation is the penultimate step in the urban innovation management process, and it is important to ensure that new ideas are implemented effectively and efficiently. Here are a few ways to approach implementation in urban innovation management:

Develop an implementation plan: Create a detailed implementation plan that outlines the steps, resources, and timelines required to implement the new idea.

Build a coalition of support: Build a coalition of stakeholders and partners who will support and assist with the implementation of the new idea.

Secure funding: Secure the funding needed to implement the new idea, by identifying potential funding sources and developing a funding strategy.

Communicate with stakeholders: Communicate with stakeholders and the public to ensure that they are informed and engaged in the implementation process.

Train and educate: Train and educate city staff and other stakeholders on the new idea and how to implement it effectively.

By using these methods, cities can ensure that new ideas are implemented effectively and efficiently, and that the solutions developed are inclusive, responsive and beneficial to the community. By involving stakeholders, securing funding, monitoring, evaluating and improving the new idea, cities can develop sustainable solutions over the long term.

Continual Improvement

Continual improvement is an ongoing process that is essential in the urban innovation management process, as it allows cities to continuously evaluate and improve new ideas after they have been implemented. Here are a few ways to approach continual improvement in urban innovation management:

Monitor and evaluate: Continuously monitor the implementation of the new idea and evaluate its effectiveness, using data and feedback from stakeholders.

Feedback from stakeholders: Gather feedback from stakeholders to continuously improve the new idea.

Adapt and improve: Continuously make adjustments and improvements to the new idea, based on feedback and new information.

Measure and report: Measure and report on the performance of the new idea, and use data to identify opportunities for improvement.

Learn and share: Share learning with other cities and stakeholders, and learn from other cities and sectors to identify opportunities for improvement.

Scalability: Continuously evaluate the scalability of the new idea, and replicate successful solutions in other areas of the city or region.

By using these methods, cities can continuously evaluate and improve new ideas after they have been implemented and ensure that the solutions developed are sustainable over the long term. Furthermore, by gathering feedback, learning and sharing with other cities and stakeholders, cities can continuously improve the solutions and ensure that they are inclusive, responsive and beneficial to the community.

Stakeholder Engagement

Stakeholder engagement is an essential step in the urban innovation management process, as it allows cities to involve a wide range of stakeholders in the development, testing and implementation of new ideas.

Here are a few ways to approach stakeholder engagement in urban innovation management:

Identify key stakeholders: Identify the key stakeholders who will be affected by the new idea, and who can contribute to its development and implementation.

Develop a stakeholder engagement plan: Develop a stakeholder engagement plan that outlines the methods and tools that will be used to engage stakeholders.

Communicate effectively: Communicate effectively with stakeholders, using a variety of methods such as newsletters, social media, and workshops.

Create opportunities for participation: Create opportunities for stakeholders to participate in the development and testing of the new idea, such as through focus groups, co-creation sessions and user testing.

Listen and respond: Listen to the feedback and concerns of stakeholders and respond to them timely and effectively.

Foster collaboration: Foster collaboration among stakeholders, by bringing together different groups and organizations to work on the new idea.

Keep stakeholders informed: Keep stakeholders informed throughout the process by providing regular updates on the progress and status of the new idea.

By using these methods, cities can effectively involve a wide range of stakeholders in the development, testing and implementation of new ideas. Further, by communicating effectively, creating opportunities for participation, listening to and responding to feedback, fostering collaboration and keeping stakeholders informed, cities can ensure that the solutions developed are inclusive, responsive and beneficial to the community.

Urban innovation management can be a complex and challenging process, but it is critical for creating livable, sustainable, and resilient cities. By involving a wide range of stakeholders and taking a systematic and iterative approach, cities can increase the chances of success in implementing urban innovations.

Key Drivers Of Urban Innovations

Urban innovation is driven by a complex and constantly evolving set of factors, including demographic changes, economic pressures, environmental challenges, technological advancements, and social and political context. These key drivers are discussed below:

Demographic changes

As cities continue to grow and urbanize, the changing needs and preferences of residents are driving the development of new solutions and technologies.

Economic pressures

Cities are under increasing pressure to attract investment and grow their economies, which is driving innovation in areas such as transportation, housing, and urban design.

Climate change and sustainability

Cities are facing growing pressure to reduce their environmental footprint and adapt to the impacts of climate change, which is driving innovation in areas such as renewable energy, green infrastructure, and sustainable transportation.

Technology

Advances in technology, such as the internet of things, big data, and artificial intelligence, are enabling new forms of urban innovation and allowing cities to collect and analyze data in ways that were previously impossible.

Social and political context

Cities are facing a wide range of social and political pressures, such as inequality, poverty, and political instability, which are driving innovation in areas such as social services and citizen engagement.

International cooperation

Cities are also looking to other cities for inspiration and cooperation, through international networks, sharing of best practices and learning from each other to bring new ideas and solutions to their own cities.

All these drivers are interrelated and constantly evolving, shaping the future of the cities and the ways we live in them. Depending on the context, more drivers maybe considered.

Implementation Barriers For Urban Innovations

Urban innovation is critical to creating livable, sustainable and resilient cities. However, the implementation of urban innovations can be challenging, due to a variety of barriers that can impede progress. These barriers can include a lack of funding, bureaucratic red tape, lack of political will, lack of technical expertise, resistance to change, lack of data, inadequate infrastructure, short-term thinking, inadequate regulation, and lack of community engagement and more. Some of the common barriers include:

Funding

Urban innovations often require significant financial resources to implement and maintain, and many cities may not have the necessary funding to support these initiatives.

Bureaucratic red tape

Cities often have complex bureaucratic systems that can make it difficult to implement new initiatives and programs. This can include long and complicated procedures for obtaining approvals and permits, as well as resistance from entrenched interests that may benefit from the status quo.

Lack of political will

Urban innovations may require political support and leadership to be successful, and a lack of political will can be a significant barrier to implementation.

Lack of technical expertise

Cities may lack the necessary technical expertise to implement new technologies and innovations, which can make it difficult to deploy new solutions effectively.

Resistance to change

Urban innovations may require changes in behavior and attitudes, and citizens and other stakeholders may be resistant to these changes.

Lack of data

Many cities lack the necessary data to inform and support urban innovations, which can make it difficult to identify problems and opportunities, and to evaluate the effectiveness of different solutions.

Inadequate infrastructure

Some urban innovations require new or upgraded infrastructure, such as transportation systems, water and sewage systems, and energy systems, which can be costly and time-consuming to implement.

Short-term thinking

Cities often face pressure to produce short-term results, which can make it difficult to invest in long-term urban innovations that may take years to bear fruit.

Inadequate regulation

Cities may lack the necessary regulations and policies to support urban innovations, which can make it difficult to implement new solutions effectively.

Lack of community engagement

Urban innovations often require community engagement and participation to be successful, and a lack of engagement can make it difficult to implement new solutions effectively.

These barriers can make it difficult for cities to implement new technologies, policies and programs that can improve the quality of life for residents and make urban environments more sustainable and resilient. Understanding these barriers and finding ways to overcome them is crucial for the successful implementation of urban innovations and the creation of more livable and sustainable cities.

3

LEARNINGS FROM GLOBAL CASE STUDIES

Cities are not just physical spaces, they are the collective dreams and aspirations of their inhabitants brought to life.

- Jane Jacobs

Urban innovations are new and creative solutions that aim to address the complex challenges facing cities today. From improving transportation and housing to promoting sustainability and social inclusion, urban innovations have the potential to transform the way cities operate and improve the lives of residents.

In recent years, cities around the world have been experimenting with a wide range of urban innovations, and many of these initiatives have been successful in addressing local challenges and improving the livability and sustainability of cities. Global case studies on urban innovations provide valuable insights into the potential of different approaches to urban innovation and can serve as inspiration for other cities looking to implement similar initiatives.

These case studies cover a wide range of urban innovations including transportation, housing, energy, and digital technologies, and provide examples of how these innovations have been implemented in different cities around the world. They also provide information on the challenges and opportunities faced by cities in implementing these innovations, and the impact that these innovations have had on the lives of residents. Let us look at some of these urban innovations.

New York: High Line Park

One example of urban innovation in New York City is the High Line, an elevated park built on an abandoned railroad track that runs through the West Side of Manhattan. The High Line was developed as a public-private partnership between the City of New York and the Friends of the High Line, a community-based non-profit organization. The park was built on a disused elevated railway that had been out of service since 1980, and it was designed to provide a new public space while also addressing environmental and social challenges.

The High Line has become a popular destination for visitors and residents alike, attracting millions of visitors each year and providing a unique and popular green space in a densely urban area. The park also serves as an example of sustainable design, featuring native plants, energy-efficient lighting, and a rainwater harvesting system. Additionally, the High Line has played a key role in revitalizing the surrounding neighborhoods, attracting new businesses and residents to the area, and driving economic development.

The High Line has transformed a disused railway into a valuable public space, and it has also played a key role in revitalizing the surrounding neighborhoods. It's a great example of how cities can turn their underutilized spaces into community assets, while also addressing environmental and social challenges. The park has become a popular destination for visitors and residents alike, attracting millions of visitors each year and providing a unique and popular green space in a densely urban area.

London: Cycle Superhighway

The London Superhighway project, also known as the Cycle Superhighway, was launched in 2010 by the Mayor of London with the goal of creating a network of safe and direct routes for cyclists throughout the city. The project consists of a series of dedicated cycle lanes that are physically separated from motor traffic, providing a safe and efficient infrastructure for cyclists.

The London Superhighway has had a significant impact on cycling in the city, making it safer and more convenient for people to cycle. The dedicated cycle lanes have also helped to reduce the number of cycling accidents and fatalities. Furthermore, the project has also played a role in reducing air pollution and traffic congestion.

The London Superhighway has transformed the city by creating a safe and efficient infrastructure for cyclists. The project was designed to improve the accessibility of the city and to encourage more people to cycle, thus reducing the number of cars on the road and helping to reduce congestion, pollution, and emissions. Additionally, the project has also played a key role in promoting active transportation, a healthy lifestyle and a more sustainable city.

Bogota: 100 in 1 Day

The "100 in 1 Day" festival in Bogotá, Colombia is an

annual event that encourages citizens to produce small-scale, temporary interventions in public spaces that aim to improve the livability of the city. The festival is organized by the local government, in partnership with community organizations and local businesses.

The festival allows citizens to propose and implement their own ideas for public space improvements, such as creating community gardens, installing public seating, and organizing cultural events. Participants are provided with a small budget and support from the local government to implement their projects.

The "100 in 1 Day" festival empowers citizens to take an active role in shaping the city and improving their communities. It also serves as a platform for citizens to express their creativity and to test new solutions for urban challenges. Additionally, the festival also helps to build community and a sense of ownership among citizens towards their city. Furthermore, the festival has also played a key role in promoting civic engagement, social inclusion, and community building.

The "100 in 1 Day" festival showcases how cities can involve citizens in the innovation process, by providing them with the tools and resources to take an active role in shaping their communities and cities. It's a unique way of promoting citizen engagement, encouraging creativity and testing new solutions for urban challenges in a low-risk and low-cost way.

Amsterdam: Smart City Platform

The Amsterdam Smart City platform, developed by the city of Amsterdam, Netherlands, is a digital platform that uses data and technology to improve the efficiency and sustainability of urban services such as transportation, energy, and waste management. The platform uses data from various sources such as sensors, cameras, and other devices to collect data on the city, and then uses this data to optimize the performance of urban services.

One of the key features of the platform is the "Smart Traffic" application, which uses data from cameras and sensors to monitor traffic flow in real-time, and adjusts traffic signals and routing to reduce congestion and emissions. The platform also includes a "Smart Lighting" application, which uses sensors and cameras to monitor the brightness of street lights and adjust the lighting levels to reduce energy consumption. Furthermore, the platform also includes an

"Air Quality" application, which uses sensor data to monitor air quality in different areas of the city, and provide real-time information to citizens and city officials.

The Amsterdam Smart City platform has helped to improve the efficiency and sustainability of urban services, and has also played a key role in reducing congestion, emissions and energy consumption. Additionally, the platform also provides citizens with real-time information on various aspects of the city, such as air quality, traffic and weather, thus empowering them to make informed decisions. Furthermore, the platform also serves as a platform for innovation and experimentation, as it allows the city to test new solutions and services in a controlled and scalable way.

Stockholm: Congestion Pricing Policy

An example of policy innovation is the "congestion pricing" policy of Stockholm, Sweden. In 2007, Stockholm introduced a congestion pricing system, which charges drivers a fee for entering the city center during peak hours. The policy aims to reduce traffic congestion and improve air quality by encouraging drivers to use alternative modes of transportation such as public transport, cycling or walking.

The policy has been successful in reducing traffic congestion in the city center by around 20%, and it also helped to reduce air pollution by around 10%. Additionally, the policy has also helped to improve the efficiency of public transport, as more people have switched to using it. Furthermore, the policy also helped to reduce the number of accidents, as the decrease in traffic has made the streets safer.

The congestion pricing policy in Stockholm has helped to reduce traffic congestion and improve air quality, while also providing an alternative source of revenue for the city. Additionally, the policy also helped to promote sustainable transportation and to improve the efficiency and accessibility of public transport. Furthermore, the policy also served as a model for other cities looking to reduce traffic congestion and improve air quality.

The policy innovation has also been implemented in other cities such as London, Singapore, and Milan, and it's considered a good example of how cities can use policy as a tool to address urban challenges, such as traffic congestion and air pollution, while also promoting sustainable transportation and alternative sources of revenue.

Toronto: Green Roofs Initiative

An example of nature-based solution innovation in a city is the "Green Roofs" initiative in Toronto, Canada. Green roofs are roofs that are partially or completely covered with vegetation, such as grass, shrubs, or trees. They provide a range of benefits to cities, including reducing heat island effect, improving air quality, and reducing stormwater runoff.

In Toronto, the city government has implemented a green roof initiative to encourage the development of green roofs on new and existing buildings. The initiative includes a range of financial incentives and technical support to help building owners and developers install green roofs. Additionally, the city also has a green roof bylaw which requires all new buildings over a certain size to have a green roof or to make a payment to a fund for public green roofs.

The Green Roofs initiative in Toronto has helped to increase the amount of green space in the city and improve the overall livability of the city. The green roofs provide a range of benefits such as reducing the heat island effect, improving air quality, and reducing stormwater runoff. Additionally, the initiative also helped to promote sustainable urban development, by reducing the ecological footprint of buildings and making the city more resilient to the effects of climate change.

San-Francisco: Pavement To Parks

The "Pavement to Parks" initiative in San Francisco California, USA is a program that converts underutilized or underdeveloped areas of the city into small public parks, or "parklets", by re-purposing on-street parking spaces. The initiative aims to create more public spaces for residents and visitors, improve the livability of the city, and promote sustainable urban development.

The initiative is implemented by the San Francisco Planning Department, in partnership with local businesses community groups, and other stakeholders. The process begins with a community group or business proposing a parklet location and design. The proposal is then reviewed by the Planning Department, and if approved, the community group or business is responsible for designing, funding and maintaining the parklet.

The Pavement to Parks initiative has helped to create more public spaces in the city, improve the livability of the city, and promote sustainable urban development. The parklets are designed to be simple, low-cost, and easy to install, making them a flexible solution for creating more public space in the city. Additionally, the initiative also helps to improve the city's walkability and to create a more vibrant streetscape. Furthermore, the initiative has helped to improve the overall livability of the city in a simple and low-cost way.

South Burlington: Blockchain For Land Registry

An example of blockchain technology innovation in a city is the use of blockchain for land registry in the city of South Burlington, Vermont, USA. The city's government has developed a blockchain-based land registry system that allows property owners to digitally record and transfer ownership of their properties. The system uses blockchain technology to create a tamper-proof record of property ownership, making it more secure and efficient than traditional land registry systems.

The system allows property owners to create a digital record of their property, which includes information such as the property's address, size, and ownership history. Property owners can then transfer ownership of their property by updating the digital record on the blockchain. The system also includes smart contract functionality that automates the process of transferring ownership, such as the transfer of funds and the update of the digital record.

The blockchain-based land registry system in South Burlington has helped to improve the security and efficiency of the land registry system by creating a tamper-proof record of property ownership and automating the process of transferring ownership. Additionally, the system also helps to reduce costs and increase transparency by eliminating the need for intermediaries in many transactions.

Singapore: Augmented Reality For Public Transport

An example of augmented reality (AR) innovation in a city is the use of AR for public transportation navigation in the city of Singapore. The city's public transportation operator, Land Transport Authority (LTA), has developed an AR-based mobile app that helps commuters navigate the city's complex transportation network. The app uses AR technology to overlay information about bus and train routes, stops, and schedules on the user's live camera feed, making it easy for commuters to find the nearest bus or train stop, and plan their route.

The app also includes other features such as real-time information about the location of buses and trains, and alerts for service disruptions. The app also allows users to purchase tickets and manage their transport card account, and also includes a feature that allows users to access a virtual map of the city's underground train network.

The AR-based navigation app in Singapore has helped to improve the accessibility and ease of use of the city's public transportation system. The app makes it easy for commuters to find their way around the city, and to plan their route, even if they are not familiar with the city's transportation network. Additionally, the app also helps to improve the efficiency of the transportation system by providing real-time information about the location of buses and trains and alerts for service disruptions.

Barcelona: Virtual Reality For Urban Planning

An example of virtual reality (VR) innovation in a city is the use of VR for urban planning and design in the city of Barcelona, Spain. The city's government has developed a VR tool that allows architects, urban planners, and city officials to simulate and visualize different urban design scenarios in a virtual environment. The tool allows users to explore different design options, such as building heights, street layouts, and public spaces, and to evaluate the impact of these options on the city's livability and sustainability.

The VR tool includes 3D models of the city's existing buildings, roads, and public spaces, and allows users to add, remove or modify elements in the virtual environment. The tool also includes various analytical tools that allow users to evaluate the impact of different design options on things like solar exposure, wind patterns, and views.

The VR tool in Barcelona has helped to improve the efficiency and effectiveness of the city's urban planning and design process by allowing architects, urban planners, and city officials to simulate and visualize different urban design scenarios in a virtual environment. Additionally, the tool also helps to improve the livability and sustainability of the city by allowing users to evaluate the impact of different design options on the city.

San Diego: Autonomous Delivery Robots

San Diego, California, USA is using Internet of Robotic Things (IoRT) for autonomous delivery robots in the city.

Several companies have started testing autonomous delivery robots on the sidewalks of the city, which are designed to deliver packages and food to residents in a safe and efficient manner.

The robots use a combination of sensors, cameras, and mapping technology to navigate the sidewalks, avoid obstacles and safely deliver packages to their destinations. They are also equipped with security features such as GPS tracking and remote monitoring, to ensure the safety of the packages and the public. Robots are designed to operate autonomously, with the ability to self-navigate and make decisions based on their environment.

The use of autonomous delivery robots in San Diego has helped to improve the efficiency and convenience of package delivery. The robots are able to operate 24/7 and can make deliveries quickly and efficiently, reducing the need for human drivers and delivery vehicles on the road. Additionally, the robots can also reduce traffic congestion and emissions, and help to improve the overall livability of the city.

Eindhoven: 3D Printed Affordable Green Housing

The city of Eindhoven, Netherlands in collaboration with

a local construction company, has developed a project to use 3D printing technology to build affordable housing units for low-income families.

The project involves the use of large-scale 3D printers to construct the housing units, which are printed using a combination of concrete and other sustainable materials. The 3D printing process allows for the rapid and efficient construction of the housing units, while also reducing costs and minimizing waste. The units are also designed to be energy-efficient and sustainable, with features such as solar panels and rainwater harvesting systems.

The 3D printed housing project in Eindhoven has helped to address the need for affordable housing for low-income families. 3D printing technology allows for the rapid and efficient construction of housing units, while also reducing costs and minimizing waste. Additionally, the project also promotes sustainable urban development, by building energy-efficient and sustainable housing units.

New York: Social Impact Bonds

The city of New York, USA are using social impact bonds (SIBs), a type of financial instrument that allow private investors to fund public services, with the promise of a return on their investment if predetermined social outcomes are achieved. In New York, the city's government has used SIBs to fund a number of public services, such as reducing recidivism among ex-offenders and improving educational outcomes for at-risk youth. The city's government partners with non-profit organizations and private investors to develop and implement the programs, with the investors providing the initial funding. If the programs achieve their predetermined social outcomes, the investors receive a return on their investment from the city's government.

The use of SIBs in New York allows the city to fund public services without relying on traditional forms of funding such as taxes or government funding. Additionally, SIBs also promote public-private partnerships and encourage private sector involvement in addressing social issues. Furthermore, SIBs also align the interests of the private sector with that of the public sector.

Toronto: Community Gardens

In Toronto, Canada, the city's government and local non-profit organizations have developed a network of community gardens that allow residents to grow their own food and build community connections.

The community gardens are located in under-served neighborhoods, and are designed to be accessible and inclusive to all members of the community. They are managed and maintained by volunteers from the local community, and provide a space for residents to grow fruits, vegetables and herbs. The gardens also provide opportunities for community members to learn about gardening, healthy eating and sustainable living. The community gardens have helped to address the issue of food security by providing residents with access to fresh fruits and vegetables. Additionally, the gardens also promote community building by providing a space for residents to come together and work towards a common goal. Furthermore, community gardens also promote environmental sustainability by encouraging sustainable living and reducing the environmental footprint of the food system.

Amsterdam: Participation Square

The city of Amsterdam, Netherlands has developed an online platform that allows citizens to participate in the city's decision-making process and co-create solutions to local issues.

The platform, called "Participatieplein" (Participation Square), allows citizens to submit ideas and proposals for improving their neighborhoods, and to vote and comment on the ideas of others. It also allows citizens to participate in online discussions and consultations with city officials and other stakeholders. The platform also provides a feature that allows citizens to track the progress of their proposals and to receive updates on the status of their ideas.

The Participatieplein platform in Amsterdam helps to improve the transparency and accountability of the city's government by allowing citizens to track the progress of their proposals and to receive updates on the status of their ideas. Additionally, the platform also helps to build trust and collaboration between citizens and city officials by promoting open and inclusive dialogue.

Dubai: Drones For Emergency Response

the city of Dubai, United Arab Emirates (UAE) is using drones for emergency response and search and rescue operations. The city's government has developed a program that uses drones to respond to emergency situations quickly and efficiently, such as building fires and missing person cases.

The drones are equipped with cameras, thermal imaging sensors, and other advanced sensors that allow them to quickly survey the scene of an emergency and provide real-time information to emergency responders. They are also equipped with speakers and megaphones, which can be used to communicate with people in the area, giving them instructions or information about an emergency.

Drones are able to quickly survey the scene of an emergency and provide real-time information to emergency responders, which allows them to make more informed decisions and respond more quickly to the emergency. Additionally, the drones also help to save time and resources, as they can reach areas that are difficult to access by ground vehicles.

Baltimore: Lean Six Sigma For Urban Services

Baltimore, Maryland, USA has implemented Lean Six Sigma methodologies in the city government to improve the efficiency and effectiveness of city services.

Lean Six Sigma is a methodology that uses a data-driven

ɪpproach to identify and eliminate waste and inefficiencies n processes, resulting in improved service delivery, cost savings and customer satisfaction. The program in Baltimore has been used to improve a wide range of city services, from permitting and licensing to waste management and public works. For instance, the program has been used to reduce the time it takes to process building permits from an average of)0 days to just 30 days.

The implementation of Lean Six Sigma methodologies n Baltimore has led to cost savings, improved service delivery and increased customer satisfaction. Additionally, the program has helped to foster a culture of continuous mprovement within the city government.

Seattle: Carbon Fee And Dividend Program

The city government of Seattle, Washington, USA has mplemented a program that puts a price on carbon emissions, with the revenue generated from the program being returned to residents as a dividend.

The carbon fee and dividend program works by placing a fee on carbon emissions from large emitters such as power plants and factories. The revenue generated from the fee is then distributed to residents as a dividend, which helps to offset any increased costs of goods and services resulting from the fee. The program is designed to provide an economic incentive for individuals and businesses to reduce their carbon emissions and invest in clean energy.

The implementation of the carbon fee and dividend program has helped to reduce carbon emissions while also providing economic benefits to residents. Additionally, the program helps to promote social equity by returning the revenue generated from the program to residents, who may otherwise be disproportionately affected by the costs of carbon pricing.

Portland: Living Building Challenge

The Living Building Challenge is a green building certification program in the city of Portland, Oregon, USA that sets the most advanced benchmarks for environmental performance in the built environment.

The city of Portland has adopted the Living Building Challenge as a mandatory building code for all new municipal buildings and has incentivized private developments to meet the standard as well. The certification requires buildings to meet strict standards in areas such as energy and water efficiency, materials selection, and site development. The buildings must generate as much energy as they consume over a 12-month period and must collect and treat all of their own water.

The implementation of the Living Building Challenge in Portland has helped to promote the construction of highly sustainable buildings. The certification promotes the use of energy-efficient systems and materials, and encourages the use of renewable energy sources. Additionally, the program helps to promote the health and well-being of building occupants by providing natural light and fresh air.

Toronto: Adaptive Reuse Of Distillery District

The city of Toronto, Canada has used adaptive reuse by converting an old industrial building into a mixed-use development. The building, known as the Distillery District, was once a large distillery and brewery complex. The city government worked with private developers to convert the old industrial building into a vibrant mixed-use development that includes retail shops, restaurants, cafes, art galleries, and residential units.

The adaptive reuse of the Distillery District has helped to revitalize a once-dilapidated area of the city, creating a new destination for residents and visitors. The development has helped to create jobs, boost the local economy and attract investment in the area. Additionally, the development has helped to preserve the historical character of the district and has been designed to be an environmentally sustainable development, with features such as green roofs, rainwater harvesting, and energy-efficient systems.

Amsterdam: Circular Hotspot

The city of Amsterdam, Netherlands has been actively working to transition to a circular economy, which aims to keep resources in use for as long as possible, extract the maximum value from them while in use, and then recover and regenerate products and materials at the end of each service life.

One such example is the "Circular Hotspot Amsterdam" program, which is a collaboration between the city government, private companies and citizens to accelerate the transition to a circular economy in Amsterdam. The program focuses on several key areas including food, textiles, and construction materials.

For instance, in the food sector, the program aims to reduce food waste by connecting surplus food from supermarkets and restaurants to charities and food banks, thus reducing food waste and providing food for those in need. In the textiles sector, the program aims to increase the reuse and recycling of textiles by promoting the donation and resale of used clothing, as well as the use of circular textile production methods.

The implementation of a circular economy program has helped to promote sustainable use of resources and reduce waste. The program encourages businesses, citizens and the city government to work together towards circular economy solutions, fostering collaboration and innovation. Additionally, the program helps to promote a more equitable and resilient economy by creating jobs and economic opportunities, while reducing environmental impact.

Amsterdam: Circular Procurement

The city of Amsterdam, Netherlands has adopted a circular economy strategy that aims to move away from a linear economy model (take, make, use, dispose) to a circular economy model where resources are kept in use for as long as possible, extracting the maximum value from them before recovering and regenerating products and materials at the end of each service life.

One such example is the "Circular procurement" program, which encourages the city's purchasing department to buy products and services that are circular by design, and to include circularity criteria in tenders and contracts. The program also encourages the city's suppliers to adopt circular economy principles in their products and services. Additionally, the city also promotes the sharing and renting of products, as well as the repair and maintenance of products to extend their life, instead of buying new products. The program helps to promote the use of products and services that are circular by design, and to include circularity criteria in tenders and contracts. It also encourages the city's suppliers to adopt circular economy principles in their products and services. Additionally, the program helps to create a more resilient and sustainable city by promoting the sharing and renting of products, as well as the repair and maintenance of products to extend their life, instead of buying new products.

Philadelphia: Rain Gardens

The city of Philadelphia, Pennsylvania, USA used rain gardens to innovate public open spaces. Rain gardens are landscaped areas that are specifically designed to capture and filter stormwater runoff. They are typically shallow depressions filled with native plants and grasses that are able to absorb and filter large amounts of water.

The city installed rain gardens in residential and commercial areas throughout the city. The program reduces the amount of stormwater runoff that enters the city's combined sewer system, which can lead to flooding and water pollution. The rain gardens are designed to capture and filter stormwater, reducing the amount of pollutants that enter local waterways. The implementation of rain gardens in Philadelphia has reduced the amount of pollutants that enter local waterways. Additionally, the program helps to promote biodiversity by using native plants and grasses, which also provide habitat for wildlife and pollinators.

Atlanta: Tree Canopy Analysis Program

The city of Atlanta, Georgia, USA has implemented a program to assess and track the health and coverage of its urban forest to better manage and protect this important resource.

The program, called the "Tree Canopy Analysis", uses aerial imagery and geographic information systems (GIS) technology to map and measure the city's tree canopy coverage. The analysis provides detailed information on the location, species, and condition of trees within the city, as well as information on the benefits provided by the urban forest, such as carbon sequestration, energy savings, and stormwater management.

The program helps to provide detailed information on the location, species, and condition of trees within the city. The program also helps to quantify the benefits provided by the urban forest, such as carbon sequestration, energy savings, and stormwater management. Additionally, the program helps to support the city's efforts to increase the tree canopy coverage, by identifying areas where the tree canopy is low and where new trees could be planted.

Los Angles: Cool Roofs Program

The city of Los Angeles has implemented a cool roofs program in the city of Los Angeles, California, USA. A hea island is a phenomenon where urban areas are significantly warmer than the surrounding rural areas, due to the concentration of heat-absorbing surfaces such as buildings, roads, and pavements.

The program promotes the use of cool roofs, which are roofs that are specifically designed to reflect more of the sun's energy back into the atmosphere, rather than absorbing it. The program encourages building owners to install cool roofs by offering incentives such as rebates and tax credits. The program also includes an education and outreach component to raise awareness about the benefits of cool roofs, such as energy savings and reducing the urban heat island effect.

The program helps to reduce the urban heat island effect by reducing the amount of heat absorbed by buildings. The program also helps to promote energy efficiency by reducing the amount of energy required to cool buildings. Additionally, the program helps to improve air quality by reducing the amount of heat-trapping pollutants that are emitted by air conditioning units.

Singapore: My Virtual Neighborhood

Singapore has implemented a VR planning game that

allows citizens to participate in the planning process for new developments in the city. The VR planning game, called "My Virtual Neighborhood", is a web-based platform that allows citizens to explore and provide feedback on proposed developments in the city. Users can navigate a virtual replica of the city and its neighborhoods, and can explore proposed developments in 3D, including buildings, streets, and public spaces. They can also provide feedback on the design and layout of these developments and suggest improvements. The feedback is then collected and used to inform the actual planning and design of the developments.

The VR planning game allows citizens to participate in the planning process for new developments in the city in an engaging and interactive way. The game allows citizens to explore proposed developments and provide feedback, which helps to ensure that the proposed developments meet the needs and preferences of the community. Additionally, the game helps to promote transparency and civic engagement in the planning process.

Singapore: NEWater

Singapore has implemented a decentralized wastewater

treatment system that treats and reuses wastewater at the point of generation, rather than transporting it to a central treatment plant. The system, called "NEWater", uses advanced membrane technologies to treat wastewater to a high standard, making it suitable for industrial, non-potable and potable uses. The treated wastewater is then used for a variety of purposes such as industrial processes, irrigation, and even for drinking after meeting the standards for potable water.

The NEWater system has helped to reduce the amount of water that the city needs to import from other sources. The system also helps to reduce the environmental impact of wastewater treatment by reducing the energy and resources required to transport and treat wastewater. Additionally, the system helps to promote water conservation by encouraging the reuse of wastewater rather than discharging it into the environment. Furthermore, the system serves as a model for other cities looking to reduce their dependence on imported water and promote water conservation.

San Francisco: Optical Plastic Sorting System

The city of San Francisco, California, USA has implemented

an automated plastic waste sorting system that uses advanced technology to sort and recycle plastic waste more efficiently. The system, called "Optical Plastic Sorting System", uses sensors and cameras to identify different types of plastic materials and sort them accordingly, this way the recycling process is more accurate and efficient. The system is able to sort through plastic waste at a high speed and with high accuracy, allowing the city to recycle a greater amount of plastic waste.

It has helped to reduce the environmental impact of plastic waste by reducing the amount of plastic waste that ends up in landfills or oceans.

Seoul: Pay As You Throw Program

The city of Seoul, South Korea has implemented a "Pay-As-You-Throw" (PAYT) program that encourages residents to reduce the amount of waste they generate by charging them based on the amount of waste they put out for collection. The program uses a smart bin system, where residents are provided with RFID-enabled smart bins that track the amount of waste put out for collection. The residents are charged for the waste collection according to the weight of the waste, with higher charges for larger amounts of waste. The program also provides incentives and rewards for residents who generate less waste, such as discounts on their waste collection fees or gift certificates to local retailers.

The PAYT program encourages residents to reduce the amount of waste they generate by making them more aware of the cost of waste disposal. The program helps to increase recycling rates by providing incentives and rewards for residents who generate less waste. Additionally, the program helps to reduce the environmental impact of waste by reducing the amount of waste that ends up in landfills or incinerators.

Paris: AirParif System

The city of Paris, France has implemented a real-time air quality monitoring system that uses a network of sensors to measure the levels of various pollutants in the air and make the data available to the public in real-time. The system, called "AirParif", uses a network of sensors located throughout the city to measure the levels of pollutants such as particulate matter, nitrogen oxides, and ozone. The data is then transmitted to a central hub where it is analyzed and made available to the public in real-time through a web-based platform and mobile app. The system also includes an alert system that notifies the public when air quality reaches unhealthy levels.

The AirParif system provides citizens with real-time information about the air quality in their city and enables them to take appropriate measures to protect their health. Additionally, the system helps to support the city's efforts to improve air quality by providing accurate and timely data that can be used to identify sources of pollution and develop strategies to reduce them.

Milan: Vertical Forest

The city of Milan, Italy implemented the Vertical Forest as a residential building that features a façade covered in hundreds of trees, shrubs and plants which help to absorb pollutants and improve air quality. This building, designed by architect Stefano Boeri, is not only an example of sustainability and green building design, but it also serves as a solution to improve air quality. The plants absorb CO_2 and other pollutants, while also producing oxygen and providing natural shade. The building also features a rainwater harvesting system, and solar panels to reduce energy consumption.

The project helps to improve air quality by absorbing pollutants. The building also serves as an example of sustainability and green building design, by featuring a rainwater harvesting system, and solar panels to reduce energy consumption. Additionally, the building helps to promote biodiversity by providing habitat for wildlife and pollinators.

Xian: Green Lungs

The city of Xian, China has installed several large outdoor air purifying towers that use advanced filtration technology to clean the air in heavily polluted areas. The towers, called "Green Lungs", are equipped with filters that remove particulate matter, nitrogen oxides and sulfur dioxide from the air. They also use green plants and mosses to absorb pollutants and improve the air quality. The towers are located in heavily polluted areas of the city, such as busy streets, industrial areas and parks.

The project will help to improve air quality in heavily polluted areas of the city. The towers use advanced filtration technology to remove pollutants from the air and green plants and mosses to absorb pollutants, which helps to improve the overall air quality. Additionally, the towers serve as a symbol of the city's commitment to improving air quality and provide a visible demonstration of the effectiveness of the technology.

Bogota: Play Streets

The city of Bogota, Colombia has implemented a program that temporarily closes certain residential streets to traffic for a few hours each week, creating safe spaces for children to play and engage in physical activities. The "play streets" program is based on the idea of "ciclovia", a weekly event where main roads are closed to cars and opened up for people to walk, bike, jog, dance, and play. The program is implemented in neighborhoods with limited public space and high rates of child obesity and sedentary lifestyle. By closing streets to traffic, the program creates safe and accessible spaces for children to play and engage in physical activities.

The "play streets" program helps to create safe spaces for children to play and engage in physical activities, particularly in neighborhoods with limited public space and high rates of child obesity and sedentary lifestyle. The program also helps to promote community engagement and social cohesion by encouraging residents to use the streets as shared spaces for socializing and community activities. Additionally, the program helps to reduce traffic and air pollution in the city by temporarily closing streets to cars.

Chicago: Fire Risk Model

The city of Chicago, United States has implemented a predictive fire risk model that uses data and analytics to identify high-risk areas for fires and predict when and where fires are likely to occur. The model uses a variety of data sources, such as weather conditions, building characteristics, and emergency call data, to identify areas of the city that are at a higher risk of fires. The model also uses historical fire data to predict when and where fires are likely to occur, based on patterns and trends in the data. This information is then used to deploy fire and emergency services resources more effectively, and to target fire prevention and education efforts in high-risk areas.

The fire risk model in Chicago helps to increase the efficiency and effectiveness of fire and emergency services by identifying high-risk areas for fires and predicting when and where fires are likely to occur. The model also helps to improve fire prevention and education efforts by targeting them in high-risk areas. Additionally, the model helps to reduce the number of fires and the associated damage and loss of life by providing early warning and enabling a more rapid response.

Rotterdam: Flood Warning System

The city of Rotterdam, Netherlands has implemented a comprehensive flood warning system that uses advanced technology to predict and prevent floods. The system uses a network of sensors located in strategic locations throughout the city, such as riverbanks and low-lying areas, to collect real-time data on water levels, rainfall, and other factors that can contribute to flooding. This data is then combined with weather forecasts and computer models to predict the likelihood and potential impact of flooding.

If the system predicts a flood, it will automatically activate flood barriers and other flood protection measures, such as closing floodgates, to prevent the flood from occurring. The system also uses this information to send out early warning alerts to residents and emergency services, through phone and text message alerts, to inform them of the potential flood and take appropriate action.

The flood warning system helps to prevent floods by providing early warning and enabling a rapid response. The system also helps to reduce the damage and loss of life associated with floods by activating flood protection measures automatically. Additionally, the system helps to improve the overall flood preparedness and resilience of the city by providing accurate and up-to-date information on flood risk.

4

FUTURE OF URBAN INNOVATIONS

Innovation is the ability to see change as an opportunity - not a threat.

- Steve Jobs

The future of urban innovation is expected to be shaped by a number of trends and factors, such as the increasing use of smart technology, the growing focus on sustainability, and the need for citizen engagement. Additionally, as cities become more connected and data-driven, there will be an increased focus on digitalization and the use of technology to improve city services and infrastructure.

Strategic Actions

As cities continue to grow and evolve, they are facing increasingly complex and interrelated problems that require new and innovative solutions.

To meet these challenges, city governments must be able

to identify new opportunities, generate ideas, and implement new solutions. This requires a focus on innovation and a willingness to experiment, test and iterate to achieve the desired outcomes.

Some of the ways City governments can implement innovations include ways such as:

Encourage citizen participation

By engaging citizens in the innovation process, city governments can gain valuable insights and ideas for addressing local challenges. This can be done through online platforms, public meetings, and other forms of community engagement.

Create an innovation culture

City governments can foster a culture of innovation by encouraging experimentation and risk-taking, as well as recognizing and rewarding innovation.

Develop partnerships

City governments can partner with other organizations,

including private companies, universities, and non-profits, to share resources and expertise, and to develop new solutions.

Use data and technology

City governments can leverage data and technology to improve decision-making, and to develop new solutions for urban challenges. This can include the use of sensors, IoT, big data analytics, and other digital tools.

Foster entrepreneurship

City governments can support the growth of local entrepreneurship by providing resources and support for small businesses and startups, which can lead to new jobs and economic growth.

Incorporate sustainability

City governments can incorporate sustainable practices into their operations and policies, such as using renewable energy, reducing waste, and promoting sustainable transportation options.

By implementing these strategies, city governments can

develop new solutions and improve their ability to address urban challenges, leading to more livable, sustainable and resilient cities.

Focus Sectors

The future of urban innovation in cities is likely to be shaped by a number of trends and factors, including:

Climate Action

Cities are particularly vulnerable to the impacts of climate change, such as sea level rise, extreme weather events, and heat waves, and must therefore be designed and managed to be more resilient.

Urban innovations such as green infrastructure, nature-based solutions, and low-impact development can help to mitigate the impacts of climate change by reducing heat island effects, improving air and water quality, and reducing the risk of flooding. For example, green roofs and walls can reduce the urban heat island effect, and rain gardens can help to manage stormwater and reduce the risk of flooding.

Urban innovation can also help to create more sustainable and livable cities through the use of smart technology, such as smart grids, energy-efficient buildings, and electric vehicle charging stations. These technologies can help to reduce the carbon footprint of cities, and promote the use of clean energy sources.

In addition, urban innovation can also improve the resilience of cities through the development of early warning systems, emergency management plans, and disaster response systems. For example, the use of real-time monitoring systems, such as sensor networks and remote sensing, can provide early warning of potential hazards and help to mitigate their impacts.

Overall, urban innovation can play a critical role in building urban resilience, addressing the challenges of climate change, creating more sustainable and livable cities, and promoting citizen engagement.

Citizen Engagement

Urban innovation can play a crucial role in citizen engagement by providing new ways for citizens to participate in decision-making and shaping the development of their cities. Urban innovations such as digital platforms, mobile applications, and social media, can make it easier for citizens to access information, provide feedback and communicate with city officials. This increased access to information and communication channels can lead to greater transparency, accountability and ultimately, more active citizen engagement.

Urban innovations such as participatory budgeting, co-creation, and crowdsourcing also provide citizens with opportunities to directly participate in the decision-making process and shape the development of their cities. These innovations can also lead to the empowerment of underrepresented and marginalized communities by giving them a voice in the development of their neighborhoods and cities.

Urban innovation can play a role in citizen engagement by creating more livable and sustainable cities. This can be achieved by incorporating green spaces, walkable streets, and bike lanes, which not only improve the quality of life for citizens but also promote healthy lifestyles and community engagement.

In addition, urban innovation can also improve public services and infrastructure, which can lead to increased citizen satisfaction and engagement. For example, innovations in transportation such as smart traffic lights, real-time bus tracking, and bike-sharing systems can make it easier for citizens to move around the city and access public services.

Overall, urban innovation can play a critical role in citizen engagement by providing new ways for citizens to participate in decision-making, shaping the development of their cities and improving the quality of life.

Smart Technologies

Urban innovation can play a crucial role in the integration of smart technologies and digital innovations in cities. These technologies have the potential to transform the way cities are designed, managed and experienced. They can help to create more efficient, sustainable, and livable cities by improving the delivery of public services, reducing energy consumption, and increasing mobility options.

Smart technologies such as the Internet of Things (IoT), artificial intelligence (AI), and blockchain can be used to improve the efficiency and effectiveness of city operations. For example, IoT-enabled smart streetlights can help to reduce energy consumption and lower maintenance costs, while AI-powered traffic management systems can improve traffic flow and reduce congestion.

Digital innovations such as mobile applications, social media, and digital platforms can also play a key role in urban innovation. These technologies can help to improve citizen engagement by providing new ways for citizens to access information, provide feedback, and communicate with city officials. This increased access to information and communication channels leads to greater transparency, accountability and ultimately, more active citizen engagement.

Smart transportation systems, for example, can also be integrated in urban innovation, like dynamic traffic light systems, real-time bus tracking, and bike-sharing systems, can make it easier for citizens to move around the city and access public services.

Urban innovations in smart technology and digital innovation can also play a role in addressing some of the most pressing global challenges such as climate change, poverty, and inequality. They help to find solutions to these problems and improve the sustainability of our planet.

Overall, urban innovation can play a critical role in the integration of smart technologies and digital innovations in cities, improving the delivery of public services, reducing energy consumption, and increasing mobility options, promoting citizen engagement and addressing global challenges.

* * *

By embracing these trends and factors, cities can create more sustainable, livable and resilient urban environments for their citizens. Furthermore, by involving citizens, private sectors, and other stakeholders in the development and implementation of new ideas, cities can ensure that the solutions developed are inclusive, responsive and beneficial to the community.

AFTERWORD

Urban Innovation Lab is committed to creating better cities for the future. We provide advisory services in areas of urban innovations. If you are passionate about exploring and making innovations happen in your city, please don't hesitate to get in touch with us to check how we can support you.

Email us at contact@innovateurban.com

ABOUT THE AUTHOR

Ram Khandelwal is an urban planner with over 12 years of experience in the field. He is the founder of the Urban Innovation Lab and has served as visiting faculty at SPA, New Delhi, where he taught an elective course on smart cities planning and management. He has worked as a consultant to organizations such as the World Bank, UN-Habitat, KfW Development Bank, GIZ, French Development Agency and Foreign and Commonwealth Development Office UK. He has also worked with prominent companies such as PwC, IPE Global, Yes Bank, ICF International, and S&P Crisil. He holds a Master's degree in Infrastructure Planning from Centre for Environment Planning & Technology (CEPT) University, Ahmedabad and a Bachelor's degree in Planning from School of Planning & Architecture (SPA), New Delhi.

Printed in Great Britain
by Amazon